Weekend Revival

ALSO BY ANDREW MACK

Beasts of Chase

What the River Was

Weekend Revival

LOBLOLLY PRESS ASHEVILLE, NC

Andrew Mack

LOBLOLLY PRESS

Published by Loblolly Press

loblollypress.com

Asheville, NC

Book design by Andrew Mack. Typeset in Cormorant.

Instagram: @loblolly_press

Newsletter: loblollypress.substack.com

ISBN (2026): 979-8-9943089-6-7

Printed in the United States of America

First Printing, September 2019

Second Printing, October 2019

Re-release, July 2026

Before in our lives we have all gone
down to some river or another and
spoken with those who don't often speak

We tell them about the black fumes of our dreams

roots smoldering and asleep

I looked down the steep slope of those days

a skier getting ready for a jump

I had things to say

Frank Stanford

"Fire Left by Travelers"

HERE. DRINK.

Spring starts small, at the base of the soul.
Water wells up, but sometimes the well runs dry,
and six miles back is a stream hardly deep enough
to fit the feet of the children who play in it.

I pass the stream in the morning,
see half the town sitting on the banks.
Simultaneous and rapid baptisms.

There is a crabapple tree I know.
I head that way, singing.
I sing like I've got death riding on my shoulders.
I sing like life depends on it, because I know that it does.

This weight I carry is so heavy
I trip and fall
face-first into the dirt. An old pickup truck
drives by and leaves clouds of dust so
thick I think I might choke. I spit
the grit from my mouth. Purple thistle
scrapes my cheek
I can hear the ditch weeds speak

and I remember the time the well ran dry
at the old church. The earth everywhere that summer,
dessicated. I played under the maples in the graveyard
and came across the gravekeeper, checking on his
honey hives, fanning the bees.

He asked me what I was looking for
Nothing, I said, *except maybe some water.*
He handed me an empty paper cup
saying, *Here. Drink.*

Blooms all around me now
and a honeybee hovers close by.
A crack of thunder and the bee peels off,
leaving me in the roadside ditch, headed for the hive.
Rain pours in waves,
the trees bending sideways,

and a creek begins in small tributaries.
I am in the headwaters of some great flood
so I turn my mouth to the earth
slurping water
and close my eyes to plunge deeper
into the stream

Drink

WATER CRISIS

"In October 2000, an estimated 300 million gallons of toxic waste — including heavy metals like arsenic and mercury — flowed into Martin County's river system."

Kat Lonsdorf
NPR October 3, 2018

Water is in short supply.
What's left
tastes like licorice,
methylcyclohexane, MCMH:
described as not particularly lethal
by the West Virginia American Water Co, WVAW,
only causing:
severe burning in throat,
severe eye irritation,
non-stop vomiting,
trouble breathing,
and severe skin irritation,
such as skin blistering.
Folks here buy water
from Walmart or travel

hours south
to fill five
gallon drums to the brim
from a spring
untouched
by mountaintop removal—
Scarcity
is one thing
but when the water is
made undrinkable
the community
finds ways to ward off suffocation
sucking what's left
of their world
through recycled plastic straws.

NIGHTBIRDS

1.

Nightbirds are the devil
that will come to you
unexpectedly.

The devil is busy,
especially in the nighttime.

I heard the nightbirds can
sting you with their beak,
leave you paralyzed except
you can still open your eyes.

I heard the nightbirds can
make you sweat buckets.

I heard the nightbirds can
sing to you and make you
think what they was singing
was sweet.

I heard the nightbirds can
make you wish you had
died.

2.

Nightbirds flew out my
window one morning.
I followed them
in the pouring rain.

In the pouring rain
I watched them
disappear into
the treetops.

The treetops bowed
as the wind blew and
nightbirds sang to me
I go out walkin' after midnight

out in the moonlight

justlikeweusedtodo

OFFERING

1.

The grave plots are my
inheritance
now that our family
has all departed
to be buried at
some other church.

The sound
of this empty
graveyard:
howling
Catbirds nearby
with splayed beaks,
hunting mates.

Beauty to them
is an open mouth.

2.

Stopping at graveyards
along the way
to church
on Sunday,
we identify kin;
the way a birdwatcher
identifies birds
by their beaks
the pallor of their feathers,
their songs.
How we can
believe in God if we
cannot see
him?
my mom asks.
I reply,

So much of the world
is an open mouth
we cannot see.

Rock steady,
even in her last days,
she doesn't flinch,
only turns herself
toward the window.

Close your mouth, she says,
hold your breath
when you pass
the graves.

3.

The Sunday sermon:
followed by a
hog roast, the pig
first scalded,
then impaled to roast
over a spit, the mouth
held open by a steel bar
rotating above hickory coals
and fed by deacons and
congregants.

Grease drips
over the
flames, smoke stings
the eyes.
Nearby graves
observe carefully
while the congregation
opens their mouth,
prays,
devours.

TO GUT A DEER

First, flip it.

Hang it headside down.

Bleed it, and let the jaw hang loose.

Skin it. Flay the skin from muscle in long sweet strokes.

Slide the knife along the first layer of inside tissue.

Do not breach the muscle.

Do not breach the gut or you will release the insides too quickly.

The insides will expand and you may find yourself unable to catch it.

You may find yourself unable to carry it.

When you are ready, pierce the gut but not too far.

There is a precise depth at which the deer will reveal itself to you.

Reach inside and carry the guts outward.

Hold onto the heart:

Hunters do this, the reverend will say to you.

And now take a sliver and chew it slowly.

Savor the muscle.

Rub your hands together.

Inside the gut is summer, but out here is December.

Process the meat.

Slice off silverskin and fat, and gather it up.

You may find yourself holding your breath.

Breathe, and later wash your hands.

The water runs pink, the soap foams, and it feels cool.

Rinse. Pull your hands toward your face.

Breathe, but the smell sticks.

Wash your hands again and again but the smell sticks.

Wash your hands a hundred times and breathe.

Walk back outside and watch as he scoops the eye from
the socket. He glances up at you to see if you saw him:

It's the headmeat, he says.

WEEKEND REVIVAL

I'm at the Prosperity post office when I see Sadie sweeping flies off the floor. She works the desk sometimes too. Her son, Lou, is the postman. Three people work at the post office now when there used to be just these two. Sadie looks up from sweeping and tells me, "We don't have a janitor. Until last week it was only me and Lou. A new fella come in a few weeks ago asking if we had a job for him. Now he's the mail-sorter and part-time janitor."

She says, "He promised to sing a song for the church congregation the next Sunday he's around, but that's been two Sundays ago now and he's yet to make an appearance. Still, it's exciting times. To have some new blood in town."

Prosperity Church Marks 75 Years Despite Rocky Start

The Prosperity United Methodist Church was built by hand the flooded summer of 1947. It rained every day during construction. The foundation sank the year after. The steeple tipped sideways.

The following winter, the building burned. Sadie, a longtime member of the congregation, says she saw it coming in a dream first.

"I'm not clairvoyant," she said. "But I've seen a ghost or two."

"Let me tell you about Reverend Bible. His shoulder blades get to me. I've been watching them all these years he's been preaching here. When he turns to face the cross those shoulder blades make me quiver down there. It's like he's somehow stashedheaven and hell between the blades and where they merge at the spine that's where I want to be too"

It's July and the Reverend's got a new lease on life.

"Open heart surgery can't keep me down because I know HE'S got a plan. Whatever devil is working in my heart, I tell it: "You're welcome to eat the flesh of this form—I invite you in for now. But be careful. Making a living here has never been easy. I don't imagine things will change now."

The reverend's wife said to the Herald reporter last Sunday, "It's hard to keep him still, so yes it's exciting times right now. We'll be back in the church pew this coming Sunday, and he'll be back preaching the next."

The congregation meets on
the steps of the church., facing the
sunrise. Sadie says to everyone
gathered out here this golden
Sunday, "Yesterday, noon was
the last I seen him. He saw me
smoking a cigarette and come to
join me on the porch. He asked
me for a drink of water so I said
yes, but I got him lemonade
instead. He drunk it in one
gulp and told me how good
a porch this might be for nap-
taking , and I told him *yeah* and
I went inside to drop the dishes.
When I come back out, he was gone.
I yelled down the street, *Come
back anytime!* and I meant it.
It was nice to be visited.

The Reverend climbs the
steps of the church slowly
and quiets the crowd
with his hands—
ready for a sermon
and then communion.
He says Last week
I saw a wren crash midflight
into the window, breathe deeply
then get up and fly away.
Can anyone tell me
this ain't resurrection?

The Reverend will tell you how difficult it is to make a living as a pastor in a town this small.

"I'm the pastor for four churches. Ask me how many sermons I've preached and all I know is God has a better calculator than I do."

The Reverend will tell you how difficult it can be to find a skilled musician in a town this small.

"We had a man named Earnest here for a few years. He could play piano, but only the white keys. He done real good, though."

Sadie remembers her mother well.

"She used to point at me with fingers coated in honey, and her breath smelled like nectarines. When I was young, I climbed a tree in the middle of a thunderstorm. She yelled out the window, *Get down from there, you'll get struck!* She said I ran with the thunder and lightning beneath my feet. Her breath smelled like nectarines."

At home, Sadie stacks the pots and pans the way she has since before Lou died, except now she does the drying too. She drops the dish cloth and prepares herself to go out. She asks the empty house if what she's wearing is okay. She asks the mirror about her hair. She's thinking how a mother shouldn't outlive her son, so she walks down the hall talking to the rooms. She's got on white sneakers and a blue jean dress. Small white flowers along the hem. It's been raining since Lou died, but it doesn't take her long to walk to the post office. She starts sweeping the flies again, the man who used to work here with her and Lou hasn't been seen in months. She sweeps the flies now like it's the only job she's ever known. The door chime rings and a man walks in. "Where you been?" she says. She looks up and find him looking beat up, head to toe covered in mud, dust. Blood along his arm. His clothes are ripped and he has no shoes on. "I got lost" he says. Sadie think how even this beat up he looks like Lou, so she whispers to the rain how she wishes it would rain and never stop. How she wishes it would rain and rain until this whole town flooded and there were no more men around who reminder her of Lou. Until the ground liquified and swallowed this other man in front of her whole. Mud raising up the bodies from the graveyard and she would walk out to where Lou was buried. She'd drag his body home and sit him in a chair across from her seat on the couch. She'd scoot over close to him and she'd whisper, "I missed you." Water dripping from her hair. Her blue jean dress caked in mud. Small white flowers along the hem.

The Reverend stands at the head of Lou's funeral to give his message.

"Listen. Prosperity don't come cheap, and the Devil will tell you opposite. He will tell you that fortune isn't hard won, just luck. He will tempt you with riches come quick. Lou died by lightning strike while out clear cutting his own trees. Electricity traveled down his midline, the nape of his neck, down through the arc of his spine and shot out his shoes. You can find God behind all of this. Behind the lightning, the thunder, and from fire too. Even death. You might even see death in the man who speaks in groans and whispers with crakced and jagged bones sticking up beneath a cotton shirt. You'll see death in a man who hides beneath years of skin. A man whose mouth is the oxbow creek and who tongues is a moccasin tucked behind tree-stump teeth. Death can gather us at anytime. Descened deep enough and you'll find Death at the right hand of the Lord. Descend deep enough and you'll see that's where Lou is now. So, now is time for mourning, but celebration too. Hallelujah, Hallelujah. Praise the Lord. Sadie, sing the blues! Ruth, prepare the hog. Men, go out and find berry pies. Bring the sweetwater. Play horseshoes on the front lawn. Tell each other stories. Prosperity don't come cheap, so be joyful at what you go. Be grateful, even."

Behind the Reverend, the cross stands guard.
The congregation is silent.
They know the power of grace, of revival.

LIGHTS

The woman that used to live here has now died. If she was alive, she'd invite us over to drink lemonade on the porch. It would be late afternoon, because that's when she took visitors. We'd ask her, Have you been to the cave out back and she'd say Of course not.

Jessie knows this place well. He leads the way. We take a shot of whiskey and duck inside the cave. We wade through the creek as we descend underground. Jessie smears mud on my face and arms. To camouflage you from the spirits, he says.

I lead the way through a side tunnel that starts out dry but begins to fill with water as we walk. My shoulders scrape the wall and I am crouched into a ball when the tunnel begins to descend. I slide like a waterfall just as Jessie yells Don't forget to grab the rope! and then I am launched out the end of the tunnel. I catch myself, swinging by the ancient wet rope over a black pool of water.

My headlamp flickers and dims. I hear a breath next to me and then the old woman whispers through the cave, *let go.* I plummet into the pool.

SNAKES

Last night I dreamt of snakes
in my bedroom.
under blankets,
in clothes, and
climbing the walls —

Last night was cold, and yet
I shed my covers
and ran outside,
bare feet touching
almost-morning dew,

and last night was just us,
a touch of something
quiet, and sometimes
slinking behind doors
and through bedrooms.

LOONS

Loons are silent in winter, but in early spring
they sing with the honeybees and skim lake-tops
with young ones attached to their backs.
Sometimes.

In spring, I peel back layers of my own winterself
to reveal who I will become in summertime,
sitting cross-legged by the fire and taking in what
warmth I can. In spring, the lake-moss becomes
carpet between my toes and the leaves unfold in ripples.

I'll go out and gather nettle and licorice root.
When spring is over, I'll remember it in the
wrinkled reds and jagged green veins of strawberry
vines. Summer will pass too and I'll remember in the
feeling of sweetness.

When fall comes, I will be brave and sleep by the
wood stove with no shoes on. I'll go on the hunt for
beavers. I'll search unsuccessfully for three days, and
then think how I might just let them be.

In fall, the loons and I come of age as the honeybee colonies collapse. Farmers will wander their orchards, shaking the branches and blossoms by hand. Some will buy honeybees from California. Traveling bees—shipped from there to Michigan, to Arkansas, to South Carolina, to Florida, then back again.

Fall will pass and I will remember it in the scent of vanilla and cigarettes. The scratch of his mouth against my skin. I will fall in love with the scarcity of time and watch while the martins begin their migration south in swarms. I'll watch while an osprey dives for easy prey along the top of the reservoir.

Soon, I will remember how loons are silent in winter, and somehow then I will know why in summer they howl.

RESERVOIR

He's wrapped up in songs,
late-summer cotton, haint-
stories, and sunbathing.
His mouth is summerworn
When he says *I love you*
the third time this hour.
From the shore of the reservoir
you teach me how the sun
sets earlier here.

How it still catches the birds
by surprise.

SYCAMORE

I want to brave those
silent parts of myself,
the silent river rushing
underneath all of the world.
I want to swim naked
in the river with you,
and brave your silence too.
From here, I smell the honeysuckle
rising off your chest,
the river moss on your lips,
and stepping closer,
now I see you are
speckled by sun and shade.
Rocks beneath my
feet slip and slide
as we climb the early days
of Spring —
Jack Pine pollen bursts
and we succumb
to the water, our
only refuge from
the dust
of the world.

LOBLOLLY PRESS

Loblolly Press is an independent press based in Asheville, North Carolina. We publish contemporary poetry, short fiction, and novels rooted in the American South.

Our work is shaped by a simple commitment: sustained editorial care, long-term partnership, and making real room for voices too often sidelined elsewhere. We seek writing with a distinctly Southern register, attentive to the pressures of living here now, and grounded in communities and experiences not always represented in traditional publishing.

We exist to expand what Southern literature can hold, and to build the kind of literary home where readers recognize themselves and feel closer to the work and to each other.

RECENT AND FORTHCOMING FROM LOBLOLLY PRESS

If Lost · Clint Bowman (2024)

Distant Relations · Cheryl Whitehead (2025)

Beasts of Chase · Andrew Mack (2025)

The Computer Room · Emma Ensley (2025)

Proud Roads · Kelly Riedesel (2025)

Preludes & Other Poems · Earl J. Wilcox (2025)

Habitats · Garrett Ashley (2026)

Of Water Never Ceasing · Kristin Entler (2026)

Slow Fire · Spencer K. M. Brown (2026)

www.ingramcontent.com/pod-product-compliance
Lightning Source LLC
LaVergne TN
LVHW050939210926
848482LV00013B/681
* 9 7 9 8 9 9 4 3 0 8 9 6 7 *